Samuel Luna de Almeida
Adelaide C Nardocci

Respiratory diseases and traffic density

Samuel Luna de Almeida
Adelaide C Nardocci

Respiratory diseases and traffic density

Spatial statistical analysis of the municipality of São Paulo, Brazil

ScienciaScripts

Imprint

Any brand names and product names mentioned in this book are subject to trademark, brand or patent protection and are trademarks or registered trademarks of their respective holders. The use of brand names, product names, common names, trade names, product descriptions etc. even without a particular marking in this work is in no way to be construed to mean that such names may be regarded as unrestricted in respect of trademark and brand protection legislation and could thus be used by anyone.

Cover image: www.ingimage.com

This book is a translation from the original published under ISBN 978-3-330-77036-2.

Publisher:
Sciencia Scripts
is a trademark of
Dodo Books Indian Ocean Ltd. and OmniScriptum S.R.L publishing group

120 High Road, East Finchley, London, N2 9ED, United Kingdom
Str. Armeneasca 28/1, office 1, Chisinau MD-2012, Republic of Moldova, Europe
Managing Directors: Ieva Konstantinova, Victoria Ursu
info@omniscriptum.com

Printed at: see last page
ISBN: 978-620-8-63079-9

Copyright © Samuel Luna de Almeida, Adelaide C Nardocci
Copyright © 2025 Dodo Books Indian Ocean Ltd. and OmniScriptum S.R.L publishing group

CONTENTS

DEDICATORY

I dedicate it to Ana Carolina de Oliveira Luna, my sweet wife.

THANKS

I would like to thank my master's advisor, Prof. Dr. Adelaide C Nardocci, and the members of the examining board, Prof. Francisco Chiavaralotti Neto and Prof. Dr. Clarice Umbelino de Freitas, for their invaluable lessons throughout this work.

I thank God, who is the Enabler of everything, my mother Eliete and my father Sebastiâo for their total support, encouragement and prayers for my journey, my dear siblings Evellin and Gu, Samir and Mel, for their absolute friendship, my grandparents Felisberto and Denacir, Joâozinho and Teresinha. To my great friends who inspire and motivate me.

PREFACE

This book is the result of a master's thesis in public health, one of the areas in which geographic science makes a major contribution. In this area I was able to realize myself, identifying a facet of the direct relationship between man and nature, pointing out the problem of health and pollution. I came to the Faculty of Public Health at the University of São Paulo (FSP-USP) through my friend Marco Antônio Rodrigues, because of my studies in geoprocessing. The work I went to do was to geocode hospitalizations for respiratory causes and vehicle counting points in the city of São Paulo. I was privileged to be able to do it in the office of Prof·Adelaide, who was coordinating the project. The privilege I refer to is because I had just met the person who would become my exceptional supervisor! At the end of the geocoding work, she showed me the challenge that lay ahead: "*These geocoded data can't be analyzed as tables in classical statistics, you who are a geographer, let's do a project to analyze them with spatial statistics? This is a new and promising area in epidemiological studies*." With great enthusiasm in the face of the many lessons I had learned in that room, I certainly accepted the challenge.

So we started a master's research project, which - defended in 2013 - provides solid input for mobility and health policies for the largest city in the southern hemisphere. The results of this work are now in your hands, reader. We hope it will be an enriching read.

The inspiration for the research and the publication of the book comes from the assessment that the risks to the population's health associated with exposure to pollutants from vehicles is an important challenge for researchers, public policy makers and all those interested in health and the environment. As a way of dealing with this issue, in its spatial approach, we used data on vehicle traffic and hospital admissions for respiratory causes in the city of São Paulo.

The data used was the geocoding by residential address of hospitalized patients - between 2004 and 2006, in the public and private system - under the age of 5 diagnosed with acute lower respiratory tract infections and chronic lower respiratory tract diseases; as well as hospitalized patients over the age of 64 diagnosed with chronic lower respiratory tract diseases. In order to carry out the spatial statistics, the municipal area was divided into a grid of cells measuring 500 meters by 500 meters, containing the traffic density values for each cell. Finally, population and socio-economic variables from the cells were incorporated into the analysis as a form of statistical control.

The databases were assembled using ArcGIS ArcInfo 9.3. Cluster analysis was carried out using the discrete Poisson model to calculate the expected risk for each age group, using the SaTScan™ v8.0 software. To study the spatial dependence between the rate of hospitalization for respiratory causes in each subgroup and the total traffic density, Moran's index (*I*) and the *Local Indicator for Spatial Autocorrelation* (LISA) were used, using the OpenGeoDa 1.2.0 *software.* The spatial regression analysis between the hospitalization rate in each group and the traffic density was carried out using the "R" Core Team package (2012).

As a conclusion of the study, it can be said that the analyses carried out found a significant spatial association between the risk of hospitalization for respiratory diseases in children under 5 and traffic density in the municipality of São Paulo. The hospitalization of children and the elderly, however, showed different spatial patterns. The autocorrelation analysis (Moran's I) showed a greater association between hospitalizations for respiratory diseases and vehicle density for children than for the elderly. The results of the spatial regression analysis showed a positive association between the rate of hospitalizations and traffic density, when controlled by the Municipal Human Development Index (M-HDI). In the case of the elderly, the regression coefficient was negative.

Finally, it can be said that traffic-related pollution is an important risk factor for children's health in the city of São Paulo. It is therefore important to say that measures to reduce exposure to pollution, as well as to reduce socio-economic vulnerability factors, should be prioritized in the formulation of policies.

1 INTRODUCTION

Assessing and managing the risks to the health of the population of large cities from exposure to atmospheric pollutants is still a challenge for researchers and policymakers in the field of health and the environment.

The efforts and measures to control industrial sources of air pollution undertaken in the 1970s and 1980s reduced emissions from these sources. On the other hand, the substantial increase in the vehicle fleet resulting from population growth, increased purchasing power and dependence on individual transportation in the world's large cities and metropolitan regions has placed vehicle emissions as a prominent source of air pollution in urban areas (Ribeiro and Cardoso, 2003; HEI, 2010).

Vehicles emit large amounts of carbon dioxide (CO2), carbon monoxide (CO), hydrocarbons (HC), nitrogen oxides (NOx), particulate matter (PM) and other substances known as toxic pollutants, such as benzene, formaldehyde, acetaldehydes, organic aromatic compounds (VOCs), etc.All these compounds, in addition to so-called secondary compounds such as ozone, nitrates and organic acids, can cause adverse effects on people's health (HEI, 2010).

The emission of pollutants by vehicles is related to the type of vehicle (light or heavy), type of fuel, age, maintenance conditions, travel speed, among others. In recent years, many regulatory and technological measures have been introduced with the aim of reducing emissions, through the definition of emission standards, the use of "cleaner" fuels and vehicle inspection programs,

Brazil was a pioneer in replacing fossil fuels with biofuels. The Pro-Alcohol Program, created in the 1970s with the initial aim of reducing the country's foreign dependence, introduced 5.6 million vehicles powered by hydrated alcohol produced from sugar cane (Novacana, 2012). In 1986, the National Environment Council created PROCONVE - PROGRAMA DE CONTROLE DA

POLUIÇÂO DO AR POR VEiCULOS AUTOMOTORES, (Brazil, 1986), which established, among other measures, maximum emission limits for new motor vehicles. In order to control vehicle emissions in the municipality of São Paulo, the Vehicle Environmental Inspection Program (CETSP, 2012) was started in 2008 as a way of controlling the emission of polluting gases released by vehicle exhausts, as is the case in many other countries. The program measures CO, CO_2 and HC levels in cars and motorcycles, and opacity and particulate matter values in diesel vehicles.

The quantification of vehicular emissions and their impact on local air quality and on people's exposure to the pollutants emitted have been the focus of major efforts by researchers around the world in recent years. Activities have involved various fronts such as carrying out emission inventories, developing emission and pollutant dispersion models, models based on land use; expanding the monitoring and data collection network, among others.

Vehicle emissions are the main source of intra-urban variation in pollutant concentrations and monitoring data generally does not capture this spatial variation. Studies examining pollutant concentration gradients around high-traffic roads show that exposure zones can vary from 50 to 1500 m apart, depending on the pollutant and meteorological conditions (HEI, 2010).

Due to the practical difficulties of measuring all the substances associated with vehicle emissions at intra-urban scales, traffic-related pollution indicators have been used to assess their contribution to air quality and human exposure. These have helped in the temporal and spatial assessment of traffic-related pollution as well as in the definition of reduction strategies.

Various indicators have been used in epidemiological studies to estimate exposure to traffic: measured or modeled concentrations of CO and NO_2 pollutants, distance from roads; geostatistical interpolation, land use regression, measures of traffic volume and density, dispersion models and hybrid models (combining data, individual monitoring and models) (Jerret et al,

2005). However, none of these can be considered ideal.

Factors such as the temporal pattern of activities, weather conditions, volume and type of vehicles, traffic speed, land use pattern, rate of chemical transformations, among others, influence the degree to which the spatial and temporal distribution of indicators reflects exposure to traffic.

A review study carried out by the *Health Effects Institute* analyzed the evidence of an association between traffic-related pollution and various health outcomes such as respiratory and cardiovascular diseases, allergies, birth-related effects and cancer. This HEI panel identified that areas between 300 and 500 m from high-traffic roads are most affected by vehicle emissions and that indicators of traffic exposure have played and will continue to play a relevant role in assessing exposure in epidemiological studies. (HEI, 2010)

The Panel also concluded that many aspects of the epidemiological and toxicological evidence of health effects associated with traffic exposure remain incomplete. However, it considered that the evidence is sufficient to support a causal link between traffic exposure and asthma exacerbation and that there is suggestive evidence of a causal relationship with asthma in children, non-asthmatic respiratory symptoms, worsening lung function, cardiovascular and total mortality, and cardiovascular morbidity. For many other health outcomes there is limited evidence of an association, but the data was considered either inadequate or insufficient to support conclusions. (HEI, 2010)

Many studies have found a strong association between asthma and other respiratory symptoms such as chronic obstructive lower airway disease, especially in children (Ko and Hui, 2012; Burr et al, 2004; Zmirou et al, 2004, Brauer et al 2007, Jerret et al. 2008, Kim et al, 2008). In adults, the role of vehicle pollution in triggering problems is more complex and has been less studied (Moding et al, 2006, Kan et al, 2007, Schikowski et al, 2008, Jacquemin et al, 2009). Some studies have also looked at the association between traffic exposure and allergies (Brauer et al, 2006, 2007; Cesaroni et al 2008,

Morgenstern et al 2008).

The greater biological vulnerability of children and the elderly to atmospheric pollution stems from physiological peculiarities. In children, factors such as a faster rate of growth, a greater area of heat loss per unit of weight, high rates of resting metabolism and oxygen consumption, allow the chemical agents present in the atmosphere to access their airways more quickly than in adults. In the elderly, factors related to low immunity and reduced ciliary function contribute to increased vulnerability to respiratory illness related to air pollutants (Gonçalves et al, 2012).

Various birth effects, such as low birth weight, small gestational age, postnatal and perinatal mortality have been associated with exposure to air pollution (Dolk et al 2000, Ritz et al, 2000).

Two studies used an average of pollutants weighted by the distance of all roads within a 250 m buffer around the births and found an association between traffic and some birth effects (Wilhem and Ritz 2003; Ponce et al 2005). Medeiros et al (2009) studied the association between traffic-related pollution and perinatal mortality in the municipality of São Paulo. This study georeferenced the information collected in a case-control study of perinatal mortality by home address and calculated the distance-weighted traffic density (DWTD). The results showed that mothers exposed to the highest quartiles of DWTD compared to the least exposed had a 50% increase in the risk of perinatal mortality.

An important aspect in studying the association between health outcomes and traffic exposure is the influence of confounding factors, especially socioeconomic status (SES). Studies on the social distribution of exposure to air pollution have postulated that groups with low SES are exposed to higher concentrations of air pollutants and that the interaction between air pollution and social factors aggravates the health effects in this group (HEI, 2010). However, these conclusions have varied between countries. Some studies carried out in cities in England, Sweden, the Netherlands and the United States have found that the poorest population is more exposed to air pollution (Deguen and Zmirou-Navier, 2010). On the other hand, results observed in other cities in England, Italy and France showed that the better-off population was exposed to high levels of pollution (WHO, 2010).

São Paulo currently has 12 air quality monitoring stations and, considering the large area of the municipality, the data collected has little spatial representativeness (CETESB, 2008). The monitoring data does not allow for the identification of exposure gradients in the intra-urban region, especially those associated with traffic-related pollution in areas with denser roads or roads with a high volume of vehicles.

The municipality of São Paulo covers an area of 1,523 km^2 and has a population of around 11 million inhabitants. The city of São Paulo has a fleet of around 6 million vehicles and 15,000 buses that circulate daily on its 18,000 kilometers of roads (DETRAN, 2011). The distribution of roads and the volume of traffic is varied, with a high density of roads and the greatest volume of traffic in the central region of the city, as well as extensive corridors and highways that cross the urban area of the municipality, which have an intense volume of vehicles on a daily basis.

Air quality control began in 1973 with measurements of SO2 and smoke, moving to automatic monitoring in 1981. Currently, a network of 12 fixed stations monitor concentrations of PM10, SO2, NO, NOx, CO and O3. Measurements of $PM_{2.5}$ concentrations began in 1999 and are still carried out manually at 3 stations in the city.

Preliminary results from the FAPESP PPP-SUS project - 2006/61616-5, found an association between traffic density in the municipality of São Paulo and hospitalizations for respiratory diseases in children under 5 and elderly people aged 65 and over, as well as cardiovascular diseases, with a clear dose response gradient (Cardoso, 2010).

A recent study carried out in the city of São Paulo found a modest correlation (r^2 =0.28) between mortality from cardiovascular diseases and indicators of exposure to traffic pollution. It used road density and traffic volume by city district as indicators of exposure, and spatial regression analysis of road density showed an association with mortality from cardiovascular diseases, but found no association with traffic volume (Habermann and Gouveia, 2012).

Toledo (2010) used a model to simulate the emission and dispersion of vehicle pollutants in the central region of the city of São Paulo, and found that increased concentrations of traffic-related pollutants (CO, NOx and PM10) were associated with a greater chance of hospitalization for respiratory diseases in children. According to Toledo (2010), children living in census tracts with CO concentrations in the 4th quartile were 80% more likely to be hospitalized for respiratory diseases.

Thus, São Paulo is one of the largest and most polluted cities in the world, and there are still few studies aimed at evaluating the intra-urban variations in traffic-related air pollution and the impact on people's health. Air pollution monitoring data is insufficient and inadequate for studying the intra-urban variability of the population's exposure to vehicle pollutants.

On the other hand, in recent years, the evolution of tools related to Geographic Information Systems (GIS) and the advancement of spatial data analysis methods have made it possible to carry out mapping studies and spatial risk estimates using small spatial units of analysis, reducing the biases of ecological fallacy resulting from the aggregation of information over large areas.

2. METHOD

The purpose of this study is to identify spatial patterns of respiratory diseases in children and the elderly and their exposure to traffic in the municipality of São Paulo.

The growing availability of computer techniques and advances in the development of Geographic Information Systems (GIS) have increased the importance of ecological studies, especially in the field of environmental health. These studies have enabled a better understanding of the temporal and spatial variation in the risk of falling ill among different population groups. Appropriate analysis tools make it possible to control possible confounding factors and avoid analysis bias.

Studies of the variability of risk at the ecological level have played an important role in understanding the social and environmental determinants of the health-disease process and introduce space as a multidimensional factor, enabling not only the surveillance of individuals at risk, but also a view of collective risk. It also seeks to identify the causes of the incidence of diseases in population groups and not the causes of the disease in the individual.

This research is based on data and information obtained from a research project funded by FAPESP (FAPESP PPP-SUS 2006/61616-5) and developed by the School of Public Health, in partnership with the Sâo Paulo State Health Department and the Sâo Paulo Municipal Health Department. The aim of the project was to correlate cases of hospital admissions in the public and private sectors with data on the volume of vehicles on the roads.

2.1 Hospital Admissions Database

Hospitalization data with the patient's home address was obtained from the Sâo Paulo State Health Department for the period from 2004 to 2007, from the public system (Autorização de Internaçâo Hospitalar - AIH) and the private system (Comunicaçâo de Internaçâo Hospitalar - CIH). However, as the CIH registration system had data up to May 2007, this study considered the period from 2004 to 2006.

To choose the diagnoses, we selected the categories according to the ICD10, whose literature points to the strongest evidence of an association with vehicle pollution. These are

For children under 5 years old:

J20 to J22 - Other acute infections of the lower airways.

J20: Acute bronchitis;

J21: Acute bronchiolitis;

J22: Unspecified acute infections of the lower airways.

J40 to J47 - Chronic lower airway diseases

J40: Bronchitis not specified as acute or chronic;

J41: Simple, mucopurulent chronic bronchitis;

J42: Chronic bronchitis, unspecified;

J43: Emphysema;

J44: Other chronic obstructive pulmonary diseases;

J45: Asthma;

J46: asthmatic status;

J47: Bronchectasis.

For elderly people aged 65 and over, ICDs J40 to J47 were selected: Chronic diseases of the lower airways.

Hospitalization data was georeferenced by the patient's home address. This procedure was approved by the Research Ethics Committee and complies with CONEP requirements. The database was acquired from the mapping and geo-information company Multispectral.

The georeferencing of hospitalizations consists of preparing the databases with the main objective of organizing, verifying and correcting the available data in order to analyze it geographically in the following stages, and involves the following steps:

a) Adjusting and improving the consistency of the tabular databases: the AIH databases are standardized (format and distribution of variables) in order to qualify their entry into the BDG. Hospitalizations with addresses identical to those registered in the National Registry of Health Establishments (Cadastro Nacional de Estabelecimentos de Saù - CNES) are excluded through *linkage* between the two databases. This procedure is necessary because of the finding, in the specialized literature and in fieldwork carried out, that when hospitalizations are registered in the system, in some cases the hospital's administrative service registers the address of the establishment itself when it does not know the patient's address, thus making it possible to continue registering the hospitalization in the database;

b) Treatment, using a specific verification algorithm, of the quality of the address and ZIP code filled in on the AIHs and CIHs through linkage with the digital street database. This procedure aims to minimize losses in the geocoding of inconsistent addresses (typing errors);

c) Both the hospitalizations recorded in the AIH and those recorded in the CIH had different data entry criteria that had to be made compatible to allow for the analysis and merging of information.

First of all, the structures of the databases are completely different. Even within the database, there have been structural changes over the years, which should be looked at to avoid losing data. Secondly, the dates of hospitalization and birth had different formats in terms of the location of the day, month and year. In txt format, they were sometimes filled in using the European model, sometimes the American model, sometimes starting with the year of the event. To correct this diversity of classifications, date correction software was built. Due to the size of the databases and the need to maintain the origin information for each one, georeferencing was done separately for AIH and CIH.

ARCGIS 9.3 software from ESRI was used for these activities, using a license acquired with funds from the FAPESP/PP- SUS 2006/61616-5 project.

Table 1 shows the general picture of all hospital admissions in the municipality of São Paulo in the period 2004-2006, as well as the number actually georeferenced for all ICD10 chapters and databases. Table 2 below shows the total number of hospital admissions in the municipality of Sâo Paulo, the number of admissions for respiratory causes and the number of admissions for the selected diagnostic categories for children under 5 and the elderly aged 65 and over.

Table 1: Total number of hospitalizations, number of geocoded hospitalizations and percentage of location by ICD-10 category and type of base (AIH and CIH) in the municipality of São Paulo, 2004-2006.

2004-2006		BASE		GEOCODED		(%)	
ICD-10	SIGLAS	AIH	CIH	AIH	CIH	AIH	CIH
Chapter I	A00-B99	60360	20931	53096	17313	87,97	82,71
Chapter II	C00-D48	94612	76836	82076	62401	86,75	81,21
Chapter III	D50-D89	10749	4517	9690	3848	90,15	85,19
Chapter IV	E00-E90	23563	24216	21729	20440	92,22	84,41
Chapter V	F00-F99	69382	5588	58229	4679	83,93	83,73
Chapter VI	G00-G99	54111	16548	44572	13936	82,37	84,22
Chapter VII	H00-H59	25396	19386	22502	17369	88,60	89,60
Chapter VIII	H60-H95	4877	4367	4187	3707	85,85	84,89
Chapter IX	I00-I99	160880	102317	147688	86162	91,80	84,21
Chap. X	J00-J99	151068	90285	138100	75627	91,42	83,76
Chapter XI	K00-K93	125240	85103	113616	70795	90,72	83,19
Chapter XII	L00-L99	25969	11300	23176	9485	89,24	83,94
Chapter XIII	M00-M99	33350	37324	28939	30801	86,77	82,52
Chapter XIV	N00-N99	76562	101328	70318	85367	91,84	84,25
Chap. XV	O00-O99	344018	120153	322434	86119	93,73	71,67
Chapter XVI	P00-P96	41815	8746	38451	7026	91,96	80,33
Chap. XVII	Q00-Q99	19315	7618	17029	6314	88,16	82,88
Chap. XVIII	R00-R99	21359	42408	19390	35397	90,78	83,47
Chapter XIX	S00-T98	139193	55002	123400	45309	88,65	82,38
Chap. XX	V01-Y98	191	1598	174	1269	91,10	79,41
Chapter XXI	Z00-Z99	29997	15350	26985	12644	89,96	82,37
Total		1512007	850921	1365781	696008	90,33	81,79

Table 2: Total number and percentage of hospitalization cases by age group in the municipality of São Paulo, 2004-2006

Period	Percentage of Total and Respiratory Hospitalizations by Age Group	
20004-2006	Total	Respiratory

< 5 years	218248 (100%)	81033 (37,0%)
> 64 years	377983 (100%)	43937 (11,6%)
All ages	2062525 (100%)	213727 (10,4%)

2.2 Traffic volume

The number of cars, motorcycles, buses and trucks was measured at 681 points in the city. The measurements were taken from Monday to Friday, at peak times in the morning (7-10 a.m.) or afternoon (5 p.m. - 8 p.m.). The number of vehicles was counted at 15-minute intervals and transformed into hourly volume. The average vehicle volume calculated for each type of road (fast traffic, arteriali, arterial2, arterial3, collector, collector 2 and local) was assigned to the other roads without measured information. All the information was entered into a georeferenced street database.

The measurement of traffic volume was carried out by technicians from the Environmental Surveillance Service of the municipality of São Paulo, students and technicians from the School of Public Health. This activity, as well as the consolidation and preparation of the database were part of the activities of the FAPESP PPP-SUS research project - 2006/61616-5 (Cardoso, 2010).

2.3 Construction of the Spatial Analysis Unit

In view of the discussions on the grouping and segregation of spatial information and the challenges of equalizing the units of analysis, a vector grid was created to apply a sample counting technique to spatial patterns called "Quadrats" (Krainski et al 2005, Lima et al 2006).

The use of a vector grid for the definition of analysis units was used as a plan for composing the uniform space of information by area (500X500), generating vector grid polygons of analysis units using the ET Geowizards 9.9 software for ArcGIS 9.3 . The use of this resource made it possible to base 2500m^2 as the spatial extent of the analysis units uniformly. The grid was built on a reference information plane, in this case the municipality's census grid (IBGE, 2000) in the UTM coordinate system, SAD 69 datum - spindle 23. In this way, the official coordinates were used for the global positioning of the grid created. They were then delimited using the "clip" tool on the municipal boundary of the sector grid base. The analysis grid was then made up of whole polygons, which were then cut out.

The sector information allocation calculation was done proportionally to the area of the grid, in order to avoid the edge effect deviations common in area calculations. It was done with the database (model input parameters) juxtaposed with the vector grid and its grid cell identifier information. The data is combined through an intersection process and the calculations are summed in the block of intersected segments and finally divided by the area of the unit of analysis.

The hospitalization rates (TI) in each area i were calculated by:

$$TI_i = \frac{E_i}{P_i}$$

E_i is the number of hospitalizations (events) in area i and P_i is the population in area i, for each age group considered. The rate of hospitalizations for respiratory diseases in children under 5 (IRR) and the rate of hospitalizations for respiratory diseases in the elderly over 64 (IRR) were calculated.

The traffic density was calculated by:

$$DT = \frac{\sum_{i=1}^{n} V_i L_i}{A}$$

Where Vi is the number of vehicles (vehicles x m/hour/m^2) and Li is the length of the i-th street segment (meters) and A is the area of the square (500mx500m). The total traffic density (DVT), the traffic density of light vehicles powered by gasoline and ethanol (DVG) (cars+ motorcycles) and diesel vehicles (buses+ trucks) (DVD) were calculated. Figure 1 shows the average volume of vehicles per hour on the road network in the city of São Paulo.

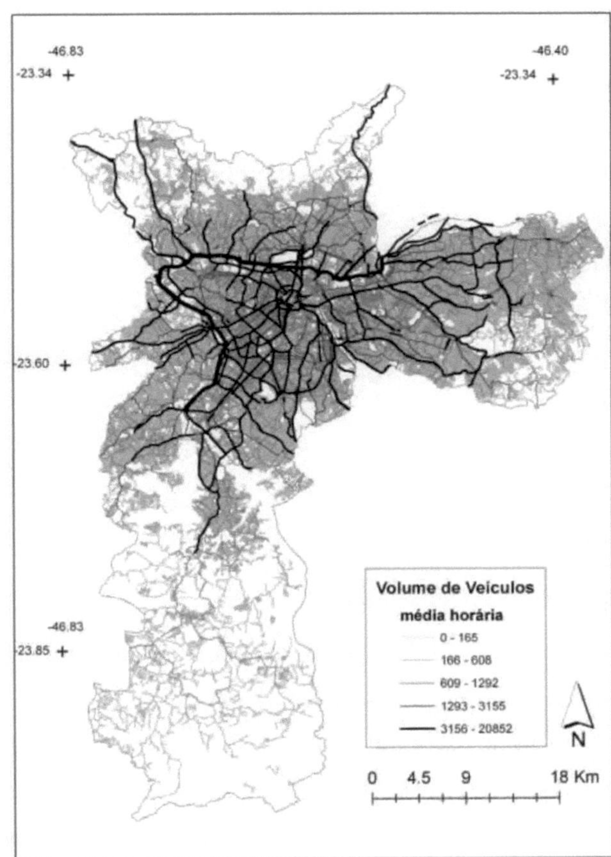

Figure 1: Average vehicle volume per hour.

3. SPATIAL ANALYSIS

According to Gatrell and Balley (1996), spatial analysis methods can be divided into three groups: visualization, exploratory data analysis and modelling. Visualization is the mapping of health events or environmental variables, and includes simple methods of spatial representation to complex map overlays. Exploratory analysis is used to summarize and describe patterns in the geographical and temporal distribution of events and the techniques employed are histograms, boxplots and cluster analysis. Spatial data modeling is used to test hypotheses or estimate relationships between variables such as the incidence of respiratory diseases and environmental or social variables (Medronho, 2009).

3.1 Exploratory data analysis.

3.1.1 Descriptive analysis

Descriptive statistics provide us with information on the median, mean, quartiles and standard deviations of the variables, allowing us to better understand the characteristics of the data, as well as the histograms which allow us to define grouping classes and, by means of the frequency, to see the value classes where the data is concentrated and the absence of a normality pattern. The correlation values between the independent variables helped us choose the control variables.

3.1.2 Area cluster analysis

For the cluster analysis, the discrete Poisson model was initially used to calculate the expected risk for the municipality, based on the number of hospitalizations for respiratory diseases in each subgroup (children and the elderly) and their respective population.

Subsequently, by means of scans with varying windows, the occurrence of risk values greater than the expected value with geographical proximity (cluster) was identified. Statistical significance was tested for 999 interactions with

$p<0.05$.

These analyses were carried out using the SaTScan™ v8.0 software (Kulldorff, 2009).

3.2 Spatial autocorrelation

To study the spatial dependence between the respiratory hospitalization rate in each subgroup and the total traffic density, Moran's index (*I*) and the *Local Indicator for Spatial Autocorrelation* (LISA) (Câmara, 2001) were used.

Moran's index (*I*) measures spatial autocorrelation from the product of deviations from the mean and is calculated by:

$$I = \frac{n \sum w_{ij}(z_i - \bar{z})(z_j - \bar{z})}{S_0 \sum_i (z_i - \bar{z})^2} \quad \text{onde: } S_0 = \sum i \neq j \ w_{ij}$$

n= number of areas;

Zi= value of the variable considered in area "i";

Z = average value of the variable in the study region;

wij= elements of the neighborhood matrix;

Zj= value of the variable considered in area "j";

Moran's I evaluates the spatial autocorrelation of the "z" variables of interest to the study in different areas *i* and *j*, (z_i, z_j), weighted by the geographical proximity measured by w / *j (Neighborhood Matrix)*, where the numerator shows the average of the products of the deviations of areas *i* and *j* in relation to the global average, and the denominator is a measure of the variability of the deviations. *Moran's I* values correspond to the slope of the regression line and, like a linear correlation coefficient, normally vary between 1 and -1, i.e.:

- *I* is *positive* when there is *spatial dependence*, with the values of the neighboring areas showing similarity to each other. A value of 1 is attributed to perfect *positive autocorrelation*;

- *I* is *negative* when there is spatial dependence, but the *values of neighboring areas are dissimilar*. The value -1 is attributed to perfect *negative autocorrelation*;

The use of a neighborhood matrix from first to third order with any contiguity (*queen* mode) was tested (Anselin L, 2005). To analyze the results, Moran's I values and the significance of the tests were compared.

The rates have been smoothed, which takes into account the average information of neighboring cells, i.e. if the gross rate of a given area shows very discrepant values in relation to the average of the neighborhood, it will be smoothed by the average of the neighboring region. If the value of the rates in a cell varies little in relation to the average of the neighboring region, its value will be little changed after smoothing. This smoothing aims to reduce the influence of random fluctuations. The value of the rate in area i is recalculated as follows:

$$\hat{\pi}_i = \frac{E_i + \sum_{j=1}^{j_i} E_i}{P_i + \sum_{j=1}^{j_i} P_i}$$

$\hat{\pi}_i$ is the smoothed value of the hospitalization rate for area i;

E_i is the number of events in area i (hospitalization cases);

E_j is the number of events in the j neighboring areas of area i;

P_i is the population in area i;

P_j is the population in the j neighboring areas of area i;

Neighborhood matrices, LISA and I'Moran analyses were carried out using OpenGeoDa 1.2.0 *software* (Anselin, 2012).

3.3 Spatial regression

Regression analysis seeks to find a good fit between the values predicted by the model and the observed values of the dependent variable in order to discover which of the explanatory variables contribute significantly to the (linear) relationship. In the case of data where spatial dependence is present, correlations may only manifest themselves if some spatial trend is taken into account.

In this way, investigating the regression residuals in search of signs of spatial structure can provide an indication of the need to use a spatial regression model. The analysis tools used so far have provided an indication that the observed values are spatially dependent.

Regression analysis on spatial data incorporates spatial dependence between the data into the model, improving the predictive power of the model. Exploratory analysis identifies the dependency structure in the data, with a view to defining how to incorporate this dependency into the regression model.

The most widespread models that capture the structure of spatial dependence are the mixed spatial autoregressive model (Spatial *Auto Regressive* = SAR or Spatial *Lag Model*) and the spatial error model (*Conditional Auto Regressive* = CAR or *Spatial Error Model*) (Anselin, 2002; Câmara et al., 2002 and Fotheringham et al., 2000).

The LAG model, indicated for this study, considers the addition of a new term in the form of a spatial relationship for the dependent variable. This regression model is given by the equation:

$$Y = \rho WY + \beta X + \varepsilon$$

being:

Y= dependent variable;

X = independent variable;

ρ= spatial autoregressive coefficient;

W = spatial neighborhood matrix or spatial weighting matrix;

β= regression coefficient;

ε= random errors with zero mean and variance σ^2; .

The null hypothesis for the non-existence of autocorrelation is that ρ= 0.

The R package R Core Team (2012) was used for the spatial regression.

4. RESULTS

The quality of the addresses in the AIHs and CIHs databases was a determining factor in not georeferencing all the cases. Even so, the georeferencing of 91.42% of AIHs and 83.76% of CIHs can be considered satisfactory.

In the period 2004-2006, hospitalizations for respiratory diseases accounted for 10% of all hospitalizations in the municipality of São Paulo. However, for children under 5, they accounted for 37% of all hospitalizations in this age group. For the elderly, total respiratory causes accounted for 11.6% of all hospitalizations.

The dependent variables in this study were hospitalization rates for respiratory diseases for the vulnerable groups considered: children under 5 and adults aged 65 and over. Total vehicle density, diesel vehicle density and gasoline vehicle density were considered as independent variables, while the HDI was used as a control variable and a household agglomeration index was also calculated, which considers the proportion of households in which 6 or more people live. However, the latter proved to be strongly correlated with the HDI, and for this reason they were not considered simultaneously in the analysis.

In the city of São Paulo, the heaviest traffic is concentrated in the central region of the city, known as the Expanded Center. This is due to the model of development and urban occupation of the city in recent decades, which concentrated infrastructure and public services in the center, while the population of workers and laborers occupied peripheral areas, with worse conditions for urban infrastructure. (Nardocci et.al, 2013)

On the other hand, the central region currently has a larger elderly population, which contrasts with the spatial distribution of the child population, which is larger in the outlying regions, where there is a younger population with lower socio-economic status, which generally has a larger number of children per family. The difference in this spatial pattern of occupation between the two age groups studied can be seen in Figure 2.

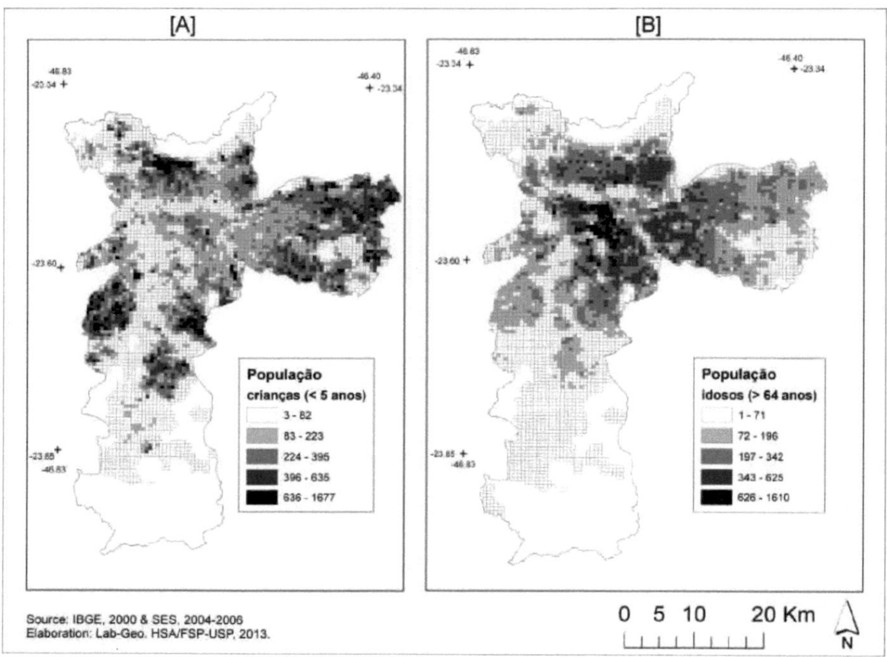

Figure 2: (A) Population of children aged < 5 years and (B) Population of adults aged over 64 years.

The concentration of roads and traffic in the central region of the city can be seen in the figures below. Figure 3(A) below shows the total traffic density (DVT) calculated by grid cell for the municipality and Figure 3(B) highlights the cells with traffic density values above the 90th percentile, i.e. the region where exposure to traffic is highest.

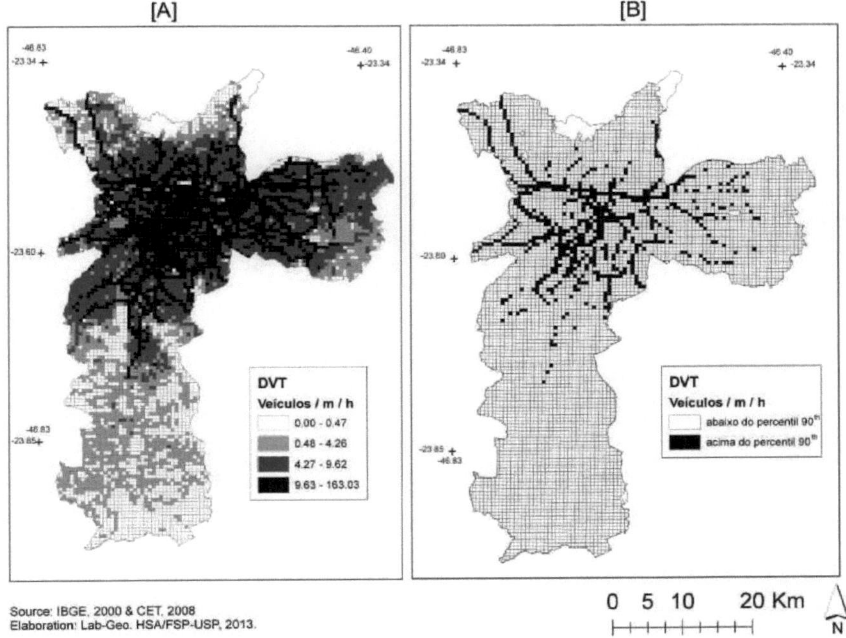

Figure 3: (A) Total traffic density (DVT) in vehicles per meter per hour, in quartiles and (B) total traffic density above and below the 90th percentile, per 500x 500m cell.

Figure 4(A) shows the values for agglomeration per household (AGLO-D), represented by the proportion of households with more than six residents, calculated using information from the IBGE census sector database. Figure 4(B) shows the MHDI values calculated by grid cell, based on the MHDI by census sector obtained from Cardoso (2010). It can be seen that in the central region of the city of São Paulo, where exposure to traffic-related pollution is greatest, the highest MHDI values and the lowest AGLO-D values are concentrated.

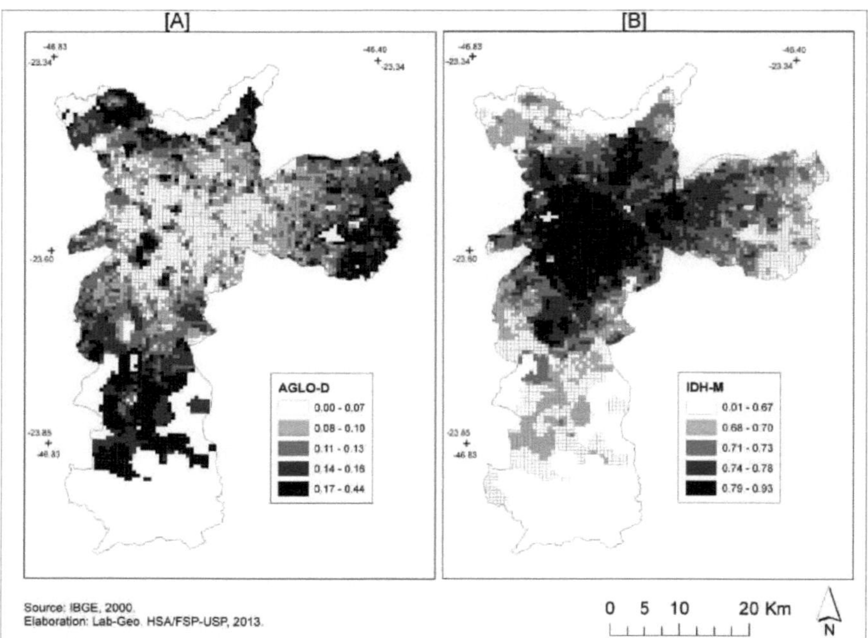

Figure 4: (A) Proportion of households with 6 or more residents (AGLO-D) (B) Human Development Index (HDI-M), by grid unit in the municipality of São Paulo.

Figures 5(A) and (B) show, respectively, the variation in rates of hospital admissions for respiratory causes per 10,000 inhabitants in children under five and in the elderly aged sixty-five and over.

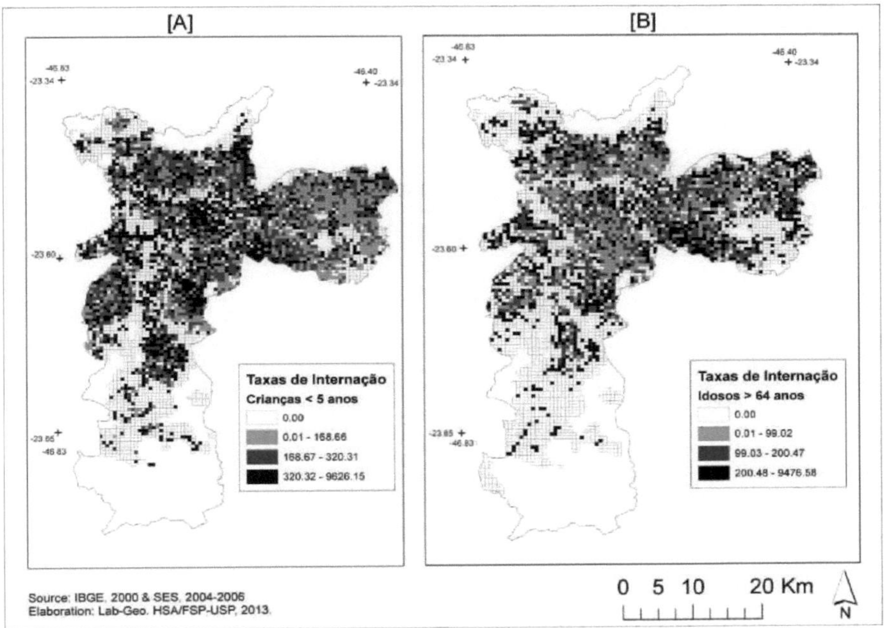

Figure 5: Hospitalization rates for respiratory diseases x 10,000 inhabitants for (A) children under 5 and (B) adults over 64 in the municipality of São Paulo, 2004-2006.

It can be seen in Figure 5(A) that for children under five, there is a concentration of high rates of hospitalization for respiratory diseases in the central region and in the areas with the highest traffic density. However, in the case of the elderly (Figure 5(B)), the spatial distribution of the highest rates is dispersed, and more present in the peripheral regions than in the central area.

Table 3 shows a summary of the descriptive parameters of the main variables considered, and it can be seen that for all variables, the distribution of values is not symmetrical, as shown in Figures 6 and 7.

Table 3: Descriptive parameters of the main variables considered in the study.

	Average	Median	Standard Deviation	Minimum	Percentage 25%	Perc.75%	Maximo
DVT (v/m/h)	11,32	6,74	15,09	0	2,24	12,79	163,03
DVG (v/m/h)	9,37	5,25	13,06	0	1,71	10,45	142,48
DVD (v/m/h)	1,96	1,38	2,12	0	0,47	2,54	20,65
HDI-M	0,72	0,71	0,07	0,02	0,67	0,76	0,93
AGLO-D	0,12	0,11	0,05	0	0,08	0,15	0,43
INCOME_R (R$)	1303,52	778,79	1530,07	9,72	498,82	1340,73	15830,00
IRRC	0,02	0,01	0,06	0	0,02	0,03	1,58
TIRI	0,017	0,004	0,069	0	0	0,015	1,447

The histograms of hospitalization rates for children and the elderly and of total and diesel vehicle traffic density are shown in Figures 9 and 10, respectively.

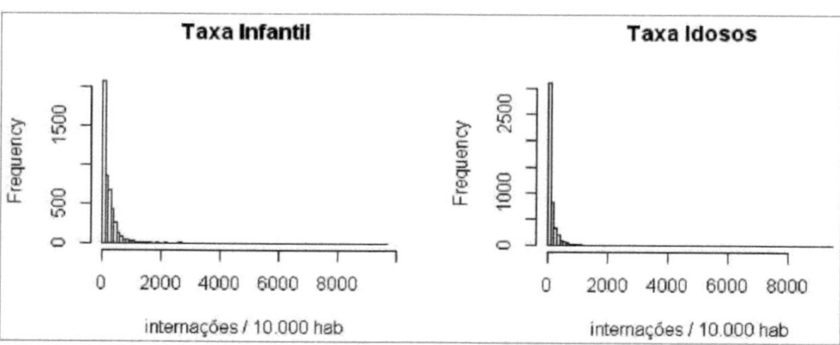

Figure 6: Histogram of hospitalization rates for respiratory diseases (A) in children under 5 and (B) in the elderly (age >64).

For both children and the elderly, there are a large number of cells with low hospitalization rates.

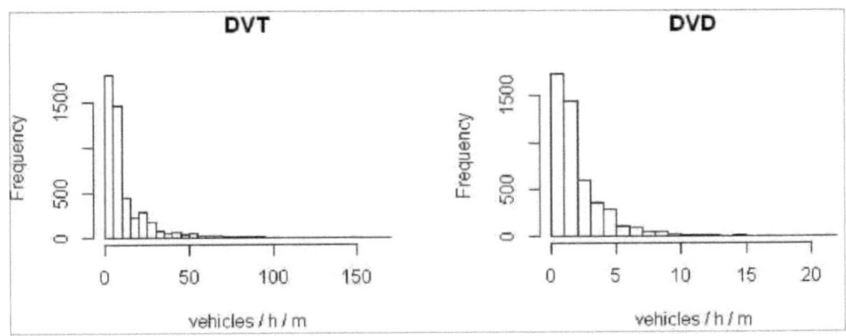

Figure 7: Histogram of traffic density values in the cells (A) total traffic density (DVT) and (B) diesel traffic density (DVD).

The correlations between the variables used in the study can be seen in Figures 8 (A) and (B), which show, respectively, the correlation graphs between the main variables for respiratory diseases in children and the elderly. The proportion of households with more than six residents (AGLO-D) is inversely correlated with the M-HDI, i.e. AGLO-D is higher where the M-HDI is lower. We therefore opted for the M-HDI as a control variable in the spatial regression models.

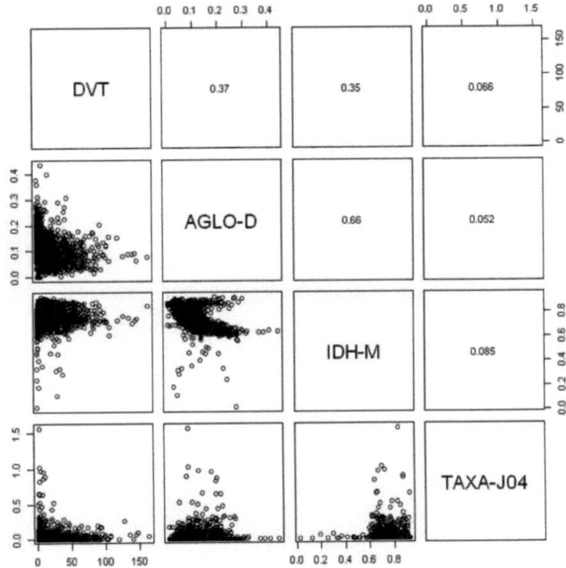

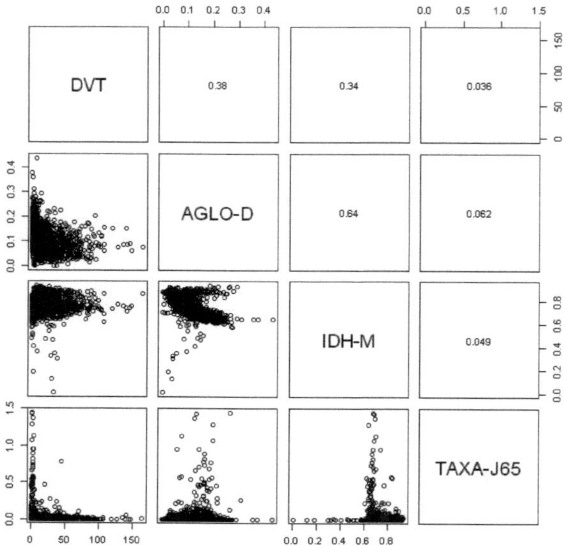

Figure 8: Linear correlation panel between the outcome variable and the independent variables used.

4.1 Cluster analysis

The aim was to identify the occurrence of spatial clusters by means of scans with varying windows, testing their statistical significance by means of a global test that considers the expected risk for the municipality, ordering them in relation to their significance. The discrete Poisson model was applied to calculate the expected rates for each spatial unit, taking into account the number of cases and the population of each cell.

Figure 9 (A) shows the clusters for respiratory diseases in children under 5 and (B) in adults aged 65 and over.

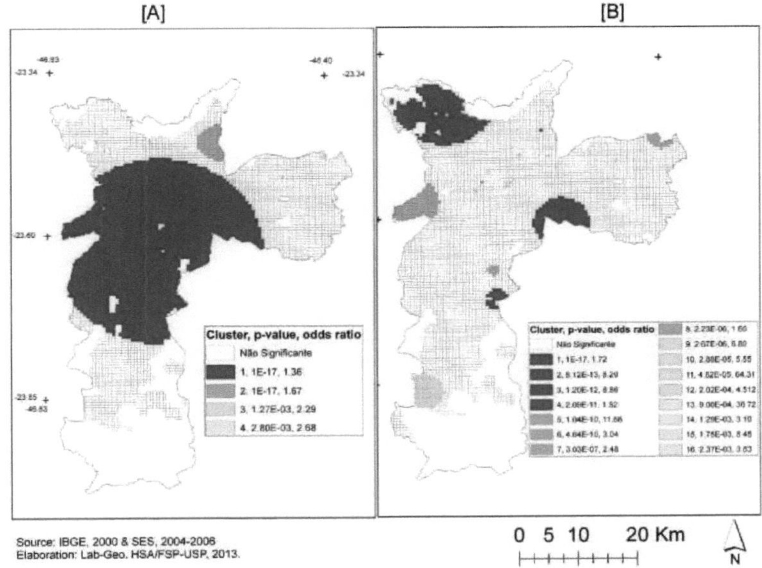

Figure 9: Identified clusters of hospitalizations due to respiratory causes

(A) children (B) the elderly.

As can be seen in the figures, the existence of clusters revealed spatial dependence through the formation of "clusters" with high hospitalization rates. In the case of children, there is a primary cluster covering the central region and the western and southern areas of the city of São Paulo. In the case of the elderly, there are small clusters outside the central region of the city.

4.2 Spatial autocorrelation

Figure 10 shows the spatial autocorrelation indices (LISA) for hospitalizations for respiratory causes in children and the elderly. For children, the correlation value was 0.5046 while for the elderly the distribution was lower and negative (-0.0363) at a significance level of $p < 0.05$.

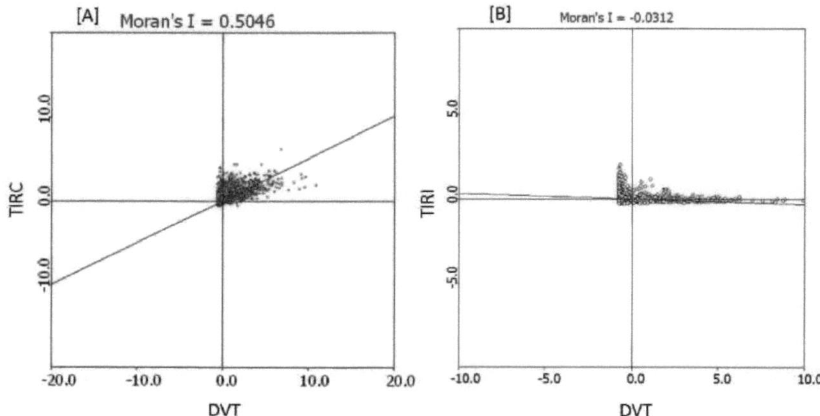

Figure 10: Moran's index of the rate of hospitalization for respiratory diseases in (A) children under 5 (B) the elderly (>64 years) and the total traffic density (DVT)

The local Moran maps obtained by weighting the 3ª order neighborhood matrix for the hospital admission rate for children and the elderly are shown in Figure 12. HIGH-HIGH areas refer to high hospitalization rates with high traffic density; HIGH-LOW refers to high hospitalization rates and low traffic density; LOW-HIGH are areas with low hospitalization rates but high traffic density and finally, LOW-LOW are places with low hospitalization rates and low traffic density.

For respiratory diseases in children, the LISA map shows HIGH-HIGH values in the central region of the city and near high-traffic roads, and is surrounded by the HIGH-LOW area.

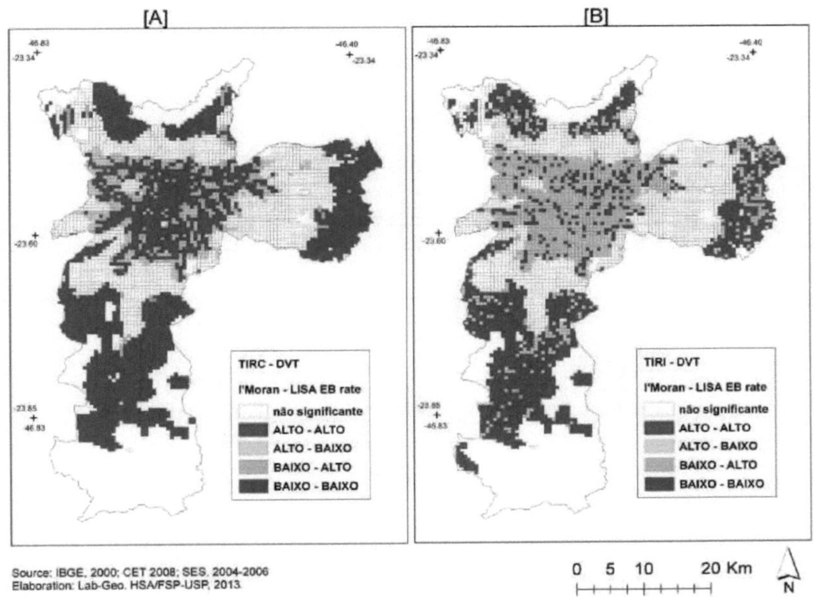

Figure 11: Moran's map of the hospitalization rate for respiratory causes in (A) children (B) the elderly and the total traffic density (DVT).

For the elderly, the HIGH-HIGH area is also found in the central region of the city, although it has a different spatial pattern, with fewer and more sparse cells. However, the rest of the central region shows a LOW-HIGH area and the HIGH-LOW area was located in the outlying regions.

4.3 Spatial Regression

For the regression analysis, the dependent variable was hospitalization rates, the independent variable was traffic density and the control variable was the M-HDI.

Table 4 shows the results of the regression analysis without taking spatial dependence into account. Table 5 shows the results of the spatial dependence test for three different neighborhood matrices: first, second and third order.

Table 4: Linear regression of the rate of hospitalization for respiratory diseases in children under 5 years of age on traffic density (DVT) and the Human Development Index (HDI-M).

Residuals:				
Min	1Q	Median	3Q	Max
0.04549	0.01995	0.01107	0.00489	1.54756
Coefficients:				
	Estimate	Std. Error	t value	Pr(> t) | |
(Intercept)	1.865e-02	8.913e-03	-2.093	0.03641 *
DVT	1.541e-04	5.767e-05	2.673	0.00755 **
HDI-M	5.736e-02	1.255e-02	4.571	4.99e-06 ***
Sign codes: 0 '***' 0.001 '**' 0.01 '*' 0.05 '.' 0.1 ' ' 1				
Residual standard error: 0.05639 on 4778 degrees of freedom				
Multiple R-squared: 0.008635, Adjusted R-squared: 0.00822				
F-statistic: 20.81 on 2 and 4778 DF, p-value: 1.005e-09				

Table 5: Comparative table of spatial dependence between neighborhood matrices for children.

Neighbour list object:	1a	2a	3a
Number of regions:	4781	4781	4781
Number of nonzero links:	35956	104672	203420
Percent. nonzero weights:	0.1573019	0.4579236	0.8899306
Average number of links:	7.520602	21.89333	42.54758
Global Moran's I for regression residuals			
Moran I statistic std. deviate	7.1432	5.788	5.4373
p-value	4.559e-13	3.562e-09	2.705e-08
alternative hypothesis	greater	greater	greater
sample estimates:			
Observed Moran's I	5.369826e-02	2.547370e-02	1.722274e-02
Expectation	-4.619755e-04	-4.280124e-04	-4.100160e-04
Variance	5.748754e-05	2.002650e-05	1.051666e-05
Lagrange multiplier diagnostics for spatial dependence			
LMerr / p-value / [df=1]	50.0045/1.534e-12	32.1022/1.463e-08	27.7186/1.403e-07
RLMerr / p-value / [df=1]	3.7592/ 0.05252	1.9497/0.1626	2.1612/0.1415
LMlag / p-value / [df=1]	51.9133/ 5.801e-13	34.3009/4.722e-09	30.954/2.642e-08
RLMlag / p-value / [df=1]	5.668/ 0.01728	4.1484/0.04167	5.3966/0.02018
SARMA / p-value / [df=2]	55.6725/0.01728	36.2506/1.344e-08	33.1152/6.443e-08

The results show that the best results were obtained for the spatial lag model (LAG) with a neighborhood matrix up to 3rd order. The results of the regression analysis with this model, for children under 5, are shown in Table 6. The results of the analysis with 1st and 2nd order neighborhoods are attached to the paper.

Table 6: Spatial regression using the Lagrange model for children under 5 and the 3ª order neighborhood matrix.

CHILDREN / LAG - 3V				
Residuals:				
Min	1Q	Median	3Q	Max
-0.0500839	-0.0194919	-0.0103451	0.0048311	1.5429566
Coefficients: (asymptotic standard errors)				
	Estimate	Std. Error	z value	Pr(> z)
(Intercept)	-1.3358e-02	8.9081e-03	-1.4996	0.133721
DVT	1.3228e-04	5.7553e-05	2.2985	0.021535
HDI-M	4.0992e-02	1.2799e-02	3.2029	0.001361
Rho: 0.27523, LR test value: 24.991, p-value: 5.7604e-07				
Asymptotic standard error: 0.052445				
z-value: 5.248, p-value: 1.5376e-07				
Wald statistic: 27.542, p-value: 1.5376e-07				
Log likelihood: 6977.572 for lag model				
ML residual variance (sigma squared): 0.0031546, (sigma: 0.056166)				
Number of observations: 4781				
Number of parameters estimated: 5				
AIC: -13945, (AIC for lm: -13922)				
LM test for residual autocorrelation				
test value: 1.8095, p-value: 0.17857				

For adults over the age of 64, the same procedure was adopted and the results obtained are shown in Tables 7, 8 and 9.

Table 7: Linear regression of the rate of hospitalization for respiratory diseases in the elderly aged over 64 and the traffic density (DVT) and the Human Development Index (HDI-M).

Residuals:				
Min	1Q	Median	3Q	Max
-0.04501	-0.01895	-0.01053	-0.00001	1.42687
Coefficients:				
	Estimate	Std. Error	t value	Pr(>|t|)
(Intercept)	4.860e-02	1.121e-02	4.336	1.48e-05 ***
DVT	-9.956e-05	7.056e-05	-1.411	0.15831
HDI-M	-4.236e-02	1.574e-02	-2.690	0.00717 **
Sign codes: 0 '***' 0.001 '**' 0.01 '*' 0.05 '.' 0.1 ' ' 1				
Residual standard error: 0.0699 on 4775 degrees of freedom				
Multiple R-squared: 0.002797, Adjusted R-squared: 0.002379				
F-statistic: 6.695 on 2 and 4775 DF, p-value: 0.001248				

Table 8: Comparison of spatial dependence between neighborhood matrices for the elderly.

Neighbour list object:	1	2	3
Number of regions:	4778	4778	4778
Number of nonzero links:	35904	104442	202952
Percent. nonzero weights:	0.1572717	0.4574914	0.8889985
Average number of links:	7.514441	21.85894	42.47635
Global Moran's I for regression residuals			
Moran I statistic std. deviate	4.6101	4.7963	5.6486
p-value	2.012e-06	8.082e-07	8.087e-09
alternative hypothesis	greater	greater	greater
sample estimates:			
Observed Moran's I	3.451940e-02	2.105128e-02	1.790236e-02
Expectation	-4.626414e-04	-4.262815e-04	-4.074627e-04
Variance	5.757994e-05	2.005222e-05	1.050708e-05
Lagrange multiplier diagnostics for spatial dependence			
LMerr / p-value / [df=1]	20.6314/5.568e-06	21.8963/2.878e-06	29.979/4.367e-08
RLMerr / p-value / [df=1]	2.0586/0.1514	5.0286/0.02493	6.0185/0.01416
LMlag / p-value / [df=1]	21.1053/ 4.347e-06	23.349/1.351e-06	32.7586/1.043e-08
RLMlag / p-value / [df=1]	2.5324/ 0.1115	6.4813/0.01090	8.798/0.003016
SARMA / p-value / [df=2]	23.1638/9.333e-06	28.3776/6.885e-07	38.7771/3.799e-09

Table 9: Spatial regression using the Lagrange model for the elderly (>64 years) and 3ª order neighborhood matrix.

ELDERLY / LAG - 3V				
Residuals:				
Min	1Q	Median	3Q	Max
-0.05858693	-0.01683152	-0.01012318	0.00024140	1.42428727
Coefficients: (asymptotic standard errors)				
	Estimate	Std. Error	z value	Pr(> z)
(Intercept)	3.5199e-02	1.1398e-02	3.0882	0.002014
DVT	-7.4741e-05	7.0345e-05	-1.0625	0.288009
HDI-M	-3.0105e-02	1.5798e-02	-1.9056	0.056699
Rho: 0.24888, LR test value: 22.883, p-value: 1.7214e-06				
Asymptotic standard error: 0.053986				
z-value: 4.6101, p-value: 4.0246e-06				
Wald statistic: 21.253, p-value: 4.0246e-06				
Log likelihood: 5946.031 for lag model				
ML residual variance (sigma squared): 0.0048512, (sigma: 0.069651)				
Number of observations: 4778				
Number of parameters estimated: 5				
AIC: -11882, (AIC for lm: -11861)				
LM test for residual autocorrelation				
test value: 30.228, p-value: 3.8419e-08				

The results of the spatial regression analysis showed a positive association between the rate of hospitalizations for respiratory diseases and traffic density and M-HDI for children. For traffic density, the increase was 1.3 hospitalizations per 10,000 children.

For the elderly, the association between the rate of hospitalization for respiratory causes and traffic density and MHDI was negative, i.e. the risk of hospitalization is greater where traffic density and MHDI are lower.

It should be emphasized that hospitalizations are serious outcomes and therefore represent only a fraction of the number of illnesses that may actually be associated with vehicle pollution.

5. DISCUSSION

This study found a significant spatial association between the risk of hospitalization for respiratory diseases in children under 5 and traffic density in the municipality of São Paulo. For the elderly, aged over 64, the results were not significant.

Cluster and spatial autocorrelation analyses reinforced the evidence of differentiated spatial patterns for children and the elderly. Among children, there was a primary cluster covering the central region and the west and south of the city of São Paulo. In the case of the elderly, there are small clusters in the peripheral regions. The autocorrelation analysis (Moran's I) also showed a greater association between hospitalizations for respiratory diseases and vehicle density for children than for the elderly.

The distribution of crude rates, in small area units, showed the presence of higher rates of hospitalization for respiratory diseases in children under five in the central region of the city, where traffic density is higher (Figure 5(A)). For the elderly, the spatial distribution of high rates was more dispersed and more present in the peripheral regions than in the central area.

The results of the spatial regression analysis showed a positive association, with a coefficient of 1.3 (Standard Error of 5.75) for hospitalizations due to respiratory diseases per 10,000 children and traffic density, when controlled for MHDI, reinforcing the results obtained in the exploratory methods. In the case of the elderly, the regression coefficient between the rate of hospitalization due to respiratory causes and traffic density, controlled for MHDI, was negative, i.e. the risk of hospitalization is greater, although not significant, where traffic density and MHDI are lower.

Comparing the results obtained in this study with others in the literature related to the respiratory system is a difficult task . The studies that take vehicle pollution into account as an indicator of exposure use more specific and restricted outcomes such as exacerbation of asthma or asthma, chronic obstructive pulmonary disease (COPD), decreased ventilatory capacity, respiratory symptoms, etc (HEI, 2010). On the other hand, the assessment of exposure to pollutants of vehicular origin is another factor that makes it difficult to compare results, as the methodologies used are very different from each other. The use of secondary data from hospital admissions restricts the standardization of the diagnosis, since it is made in different databases, by different professionals and in different places, with non-uniform criteria. With regard to children, it is also difficult to diagnose respiratory problems, especially asthma. There is no single, universally accepted set of criteria for identifying asthma, and there are several different phenotypes of the disease, which makes it almost impossible to use a single definition in epidemiological studies (HEI, 2010). This study sought to broaden the diagnoses to be considered in the analyses, selecting those that are in some way related to bronchospasm and wheezing, as an approximation of the diseases that cause bronchial reactivity in children. ICD10 J20 to J22 and J40 to J47 were selected.

The results of the analyses in children agree with the conclusions of several other studies found in the literature. A strong association between asthma and other respiratory symptoms such as chronic obstructive diseases of the lower airways, especially asthma exacerbations in children, has been observed in other studies (HEI, 2010).

Burr et al (2004) in Helsinki found that nasal congestion, eye irritation, nose irritation, rhinitis and rhinoconjunctivitis occur more among children living in the city center than in suburban areas, and attributes these findings to personal exposure to NO2, due to proximity to heavy traffic routes. The author also evaluates that these symptoms were reduced after interventions to reduce traffic in the city.

Zmirou et al, 2004, concluded that the first two years of life are a critical period for the development of asthma and that there is inconclusive evidence that gene-environment interaction makes children more susceptible if they are exposed from this age onwards. He argues that indicators of air pollutant concentrations showed a dramatic decrease in the second half of the 20th century in developed countries, with regard to pollutants of industrial origin, but that the increase in the fleet and motorized travel are contributing to the recent increase in the prevalence of asthma worldwide.

Brauer et al (2007) evaluated the development of asthmatic and allergic symptoms and respiratory infections during the first four years of life in a birth cohort study (n~ 4,000). It related concentrations of outdoor pollutants from traffic to birth registration addresses in a land use regression model. It found a correlation using logistic regression for questionnaire data on medical diagnosis of asthma, bronchitis, influenza and eczema and self-reported wheezing, dry night cough, ear infections, throat infections and skin rashes. Odds Ratio adjusted by interquartile range of pollution showed an increase for wheezing (1.2, 95%CI: 1.0-1.4), medical diagnosis of asthma (1.3, 95%CI: 1.0-1.7), throat, ear and nose infections (1.2, 95%CI: 1.0-1.3) and severe colds and flu (1.2, 95%CI: 1.0-1.4) and soot.

Jerret et al. (2008) studied a prospective cohort of 217 children aged 10-18 in Southern California to investigate associations between pollution and health. The incidence of asthma was assessed using questionnaires during 8 years of follow-up. NO2 monitors were installed outside homes for 2 weeks in summer and 2 weeks in autumn-winter and were used as a marker of traffic-related air pollution. Multilevel Cox models were used to test the association between asthma and traffic-related pollution and found a positive association with a Hazard Ratio of 1.29 (CI(5%: 1.07-1.56) between the interquartile ranges of 6.2 ppb in the annual NO2 concentration in the community. Using interquartile ranges for all measurements of 28.9 ppb annual NO2 in homes, the Risk Ratio increased to 3.25 (95%CI: 1.35-7.85), highlighting the contribution of traffic to the occurrence of asthma.

Kim et al (2008) carried out a cross-sectional study on asthma and other respiratory symptoms in children (n= 1,080) considering distances between homes and high-traffic roads in San Francisco, California, a highly urbanized region where air quality is characterized as good due to the influence of coastal breezes. Health information and residential factors were obtained through a questionnaire administered to parents. Exposure was assessed from various distance measurements from the place of residence to high-traffic roads and NOx and NO2 concentrations were also measured for a subset of families to validate the distance data. Using multivariate logistic regression analysis, an association was found between asthma and proximity to traffic.

Lee et al (2011) points to greater pulmonary effects in children and more strongly in boys than girls who attend schools where there is a concentration of pollutants from traffic in Taiwan.

For the elderly (65 and over), we chose to work only with Chronic Lower Airway Diseases, which include COPD (ICD10: J40-J47), which is more common in this age group. According to the HEI (2010), the role of vehicle pollution in triggering respiratory problems in adults is more complex, has been less studied and presents controversial results, which has led the Institute to consider that they are inadequate and insufficient to infer a causal association. Our findings suggest that there is no association between this group of diseases in the elderly and vehicle pollution.

Georeferencing by the patient's home address involved losses in the number of events, and the georeferencing efficiency was 91% for AIHs and 84% for CIHs. Although the efficiency was lower for CIHs, it can be considered acceptable for this type of study, given the large volume of existing information.

In addition to exposure to traffic-related pollution, differences in socio-economic conditions and therefore better access to health services and better housing conditions can also explain the differences observed in the spatial patterns of hospitalization rates for children and adults. In this case, the use of hospitalization data as a dependent variable should also be considered.

Hospitalizations are events that represent the most serious outcomes and may be more influenced by socioeconomic *status* and access to health services. In the case of the elderly, the proportion of older people living in the more polluted central area of the city is higher, but the majority belong to the higher socio-economic classes and therefore have more access to services and better resources for dealing with crises without the need for hospitalizations, which may not be the case for older people living on the outskirts of the city.

As for children, although those living in the central region have a better M-HDI than those living on the outskirts, they generally live in the most degraded and polluted areas.

The difference in the methods used to assess exposure to pollutants of vehicular origin is another factor that makes it difficult to compare the results of epidemiological studies on traffic pollution and health problems in the population. Exposure is rarely constant; the agents are diverse and complex; exposure also involves complex and variable behavioral aspects and people respond differently to exposure (Brauer, 2010).

Most studies of exposure variation in small areas and its association with respiratory diseases use *land* use regression models and/or models that simulate the emission and dispersion of pollutants of vehicular origin. Karr et al (2009) investigated traffic exposure and the risk of hospitalization for bronchiolitis in children in a region with low air pollution. Logistic regression analysis identified an Odds Ratio of 1.06 (95%CI: 0.92-1.17) for children living within 50 meters of high-traffic roads.

No studies were found on the association between traffic density and hospitalizations for respiratory diseases in small spatial units that used spatial analysis methods that can be directly compared to the results obtained in this study.

The city of São Paulo developed from the center to the periphery. The best quality urban infrastructure and the population with the highest socio-economic status were concentrated in the central region, while the peripheral areas were occupied by people with the worst socio-economic conditions and therefore the greatest deficit of public goods and services. This configuration has been maintained over time and has favored the concentration of traffic in the central region of the city (Nardocci et al, 2013).

The elderly population is concentrated in the central region, while the child population is larger in the peripheral regions (Figure 2). However, in the central region, there is an important difference in socio-economic status between these two age groups. While the elderly living in this area are from the higher income classes and socioeconomic conditions, the children belong to the lower classes and occupy more degraded areas, generally around high-traffic roads, viaducts and elevated highways (Nardocci et al, 2013).

Finally, it should be borne in mind that non-significant results cannot be taken as an indicator of the lack of association between exposure to vehicular pollution and health outcomes in the population, due to the methodological limitations of these studies and the city's own growth dynamics, which need to be better investigated and interpreted.

It should also be borne in mind that the conversion of socio-economic and population variables from the IBGE census sector base to 500mx500m cells may also have introduced smoothing and biases. Other advanced Bayesian inference methods for mapping risks, as well as the use of spatial units of varying sizes, could be used in future studies to deepen the analysis carried out in this work.

6. CONCLUSION

The results of this ecological study reinforce the hypothesis of a strong association between respiratory outcomes in children under 5 and exposure to vehicular pollution in the city of São Paulo. For the elderly, the study points to a greater influence of socioeconomic conditions on hospitalizations for respiratory diseases.

The use of traffic density in small spatial units as a *proxy* for exposure to traffic-related pollution was important in highlighting intra-urban gradients in exposure and hospitalization rates.

The spatial analysis methods used showed important results and significant differences in patterns between the age groups: children and the elderly. These results are relevant for subsidizing public transport policies aimed at reducing pollution in the city of São Paulo, as well as for providing subsidies for environmental health surveillance services related to air pollution.

Traffic-related pollution is an important risk factor for children's health in the city of São Paulo and measures to reduce exposure and vulnerability factors should be prioritized.

7. REFERENCES

Anselin, L. Spatial externalities, spatial multipliers and spatial econometrics. Urbana Champaign, University of Illinois, 2002.

Anselin L. Exploring spatial data with GeoDA: A Workbook. Urbana, University of Illinois, 2005.

Anselin, L. From SpaceStat to CyberGIS. International Regional Science Review, vol. 35(2), pages 131-157, April. 2012

Brazil - CONAMA RESOLUTION No. 18, of May 6, 1986, which instituted the PROGRAM FOR THE CONTROL OF AIR POLLUTION BY MOTOR VEHICLES - PROCONVE. OFFICIAL GAZETTE of June 17, 1986.

Brauer M,Gehring U, Brunekreef B, de Jongste JC, Geritsen J, Rovers M, Wichmann HE, Wijga A, Heinrich J. Traffic-related air pollution and otitis media. Environ Health Perspect, 2006,114:1414-1418

Brauer M, Hoek G, Smit HA, de Jongste JC, Geritsen J, Postma DS, kerkhof M, B. Brunekreef Air pollution and development of asthma, allergy and infections in a birth cohort; European Respiratory Journal, 2007 29: 879-888

Burr ML, Karan G, Davies B, Holmes BA, Williams KL. Effects on respiratory health of a reduction in air pollution from vehicle exhaust emissions. Occup Environ Med 2004; 61:212-218.

Câmara, G. et al Spatial analysis of geographic data. Brasilia, EMBRAPA, 2002 (ISBN: 85-7383-260-6). Available at http://www.dpi.inpe.br/gilberto/livro/analise/

Cardoso, MRA coord. project: FAPESP/PPP-SUS 2006/61616-5. Study of the relationship between respiratory and cardiovascular diseases and vehicular pollution in the metropolitan regions of the state of São Paulo with the aim of subsidizing the structuring of the activities of the health and air quality surveillance program. Technical Report, Sâo Paulo, 2010.

Cesaroni G, Badaloni C, Porta D, Forastiere F, Perucci CA. Comparison between various indices of exposure to traffic-related air pollution and their impact on respiratory health in adults. Occup Environ Med, 2008; 65:683-690.

CETESB - Environmental Sanitation Technology Company. International Seminar on Urban Air Quality Management. Available at http://www.cetesb.sp.gov.br. Accessed on 13/11/2008.

CETSP, 2012 http://www.cetsp.com.br/consultas/inspecao-veicular-environmental.aspx accessed December 2012.

DETRAN - Departamento Estadual de Trânsito de Sâo Paulo, 2011 http://detran.sp.gov.br/ accessed April 2011.

Dolk H, Pattenden S, Vrijheid M, Thakrar B, Armstrong B. perinatal and infant mortality and low birth weight among residents near cokeworks in Great Britain. Arch. Environ. Health, 2000; 55:26-30.

Fotheringham, A. S., Brundsdon, C. and Charlton, M.. Quantitative Geography: perspectives on spatial data analysis. London: Sage publications, 2000

Frùgoli Jr., H. Centralidade em Sâo Paulo: trajectórias, conflitos e negociações na metrópole. EDUSP, 2000.

Gatrell, A.C., Bailey, T.C. Interactive spatial data analysis in medical geography. Social Science & Medicine, 42 (1996), pp. 843-855

Gonçalves KS et al. (2012) The fires in the Amazon region and respiratory illness. Ciência & Saùde Coletiva, 17(6):1523-1532, 2012

Habermann M, Gouveia N. Vehicular traffic and mortality from circulatory system diseases in adult men. Revista de Saùblica, 2012; 46(1):26-33.

Health Effects Institute. Traffic-related air pollution: a critical review of the literature on emissions, exposure, and health effects. Boston, Massachusetts: HEI; May 2010. (Special Report 17).

Jacquemin B, Sunyer J, Forsberg B, Aguilera I, Briggs D, et al. home outdoor

NO2 and new onset of self-reported asthma in adults. Epidemiology, 2009; 20:119-126.

Jerret M, Arain A, Kanaroglou P, Beckerman B, Potoglou D, Sahsuvaroglu T, et al. A review and evaluation of intraurban air pollution exposure models. J Expo Anal Environ Epidemiol. 2005;15(2):185-204.

Jerret M, Shankardass K, Berhane K, Gauderman WJ, Künzli N, Avol E, et al. Traffic-related air pollution and asthma onset in children: A prospective cohort study with individual exposure measurement. Environ Health Perspect, 2008; 116:1433-1438.

Kan H, Heiss G, Rose KM, Whitsel E, Lurmann F, lonfon SJ. Traffic exposure and lung function in adults: The Atherosclerosis Risk in Communities study. Thorax, 2007; 62:873-879.

Kim JJ, Huen K, Adams S, Smorodinsky S, Hoats A, Malig B, Lipset M, Ostro B. Residential traffic and children's respiratory health. Environ Health Perspect, 2008; 116:1274-1279.

Ko FW, Hui DS. Air Pollution and COPD. Respirology. 2012;17(3):395-401

Krainski, ET; Ribeiro Jr, PJ; Neto, PRA; B, RB. An environment for monitoring citrus sudden death. VII Simpòsio Brasileiro de Geoinformàtica, Campos do Jordâo, Brazil, November 20-23, 2005, INPE, p. 223-235.

Kulldorff M. and Information Management Services, Inc. SaTScan™ v8.0: Software for the spatial and space-time scan statistics. [www.satscan.org], 2009.

Lee, YL, Wang WH, Lu CW, Lin YH, Hwang BF. Effects of ambient air pollution on pulmonary function among schoolchildren. International Journal of Hygiene and Environmental Health; 214; 2011, p 369-375

Lima, RR; Demétrio, CG Borges; Ribeiro Jr, PJ; Ridout, MS. A comparison of quadrat-based techniques for the characterization of spatial patterns in plant diseases. Revista Matematica e Estatistica, Sâo Paulo, v.24, n4, p.7-24, Oct.-

Dec. 2006.

Medeiros APP, Gouveia N, Machado RPR, Souza MR, Alencar GP et al. Traffic-related air

pollution and perinatal mortality: A case-control study. Environ Health Perspect, 2009; 7(1):127-132.

Medronho, R A Geoprocessing and health - a new approach to space in the health-disease process. Rio de Janeiro: FIOCRUZ/CICT/NECT; 2009.

Moding L, Jarvholm B, Ronnmark E, Nystrom L, Lundback B, Andersson C, Forsberg B. Vehicle exhaust exposure in an incident case-control study of adult asthma. Eur. Respir. J, 2006;28:75-81.

Morgenstern V, Zutavern A, Cyrys J, Brockow I, Koletzko S, et al..Atopic diseases, allergic sensitization, and exposure to traffic-related air pollution in children. Am J Respir Crit Care Med, 2008; 177:1331-1337.

Nardocci, A C; Freitas, C U; Oliveira, M A R; Almeida, S L; Cardoso, M R A. traffic-related air pollution and environmental inequity in Sâo Paulo, Brazil. Article Submitted to the Journal of Exposure Science and Environmental Epidemiology. 2013.

Novacana, 2012. http://www.biodieselbr.com/proalcool/pro-alcool/programa-ethanol.htm accessed December 2012.

Ponce NA, Hoggatt KJ, Wilhelm M, Ritz, B Preterm birth: the interaction of traffic-related air pollution with economic hardship in Los Angeles neighborhoods American Journal Epidemiology, 162. 2005, p 140-148

R Core Team (2012). A language and environment for statistical computing. R Foundation for Statistical Computing, Vienna, Austria. ISBN 3-900051-07-0, URL http://www.R-project.org/.

Ribeiro H, Cardoso MR. Air pollution and children's health in Sâo Paulo (1986-1998). Social Science and Medicine, v.57, p 2013-2022, 2003.

Ritz B, Yu F, Chapa G, Fruin S. Effect of air pollution on preterm birth among children born in Southern California between 1989 and 1993. Epidemiology. 2000 Sep;11(5):502-511.

Schikowski T, Sugiri D, Ranft U, Gehring U, Heinrich J, Wichmann HE, Kramer U. Contribution of smoking and air pollution exposure in urban areas to social differences in respiratory health. BMC Public Health, 2008; 8:179.

Toledo, GIFM, Evaluation of exposure to traffic-related pollution in the city of São Paulo. 2010. Thesis (PhD in Public Health) - School of Public Health, University of Sâo Paulo, Sâo Paulo.

Wilhelm M, Ritz B. Residential proximity to traffic and adverse birth outcomes in Los Angeles county, California, 1994-1996. Environ Health Perspect., 2003; 111: 1994-1996.

World Health Organization (WHO). Health Guide-lines for Vegetation Fire Events. Geneva: WHO; 1999

World Health Organization, (WHO) Environmental and health risks: a review of the influence and effects of social inequalities. Copenhagen: WHO Regional office for Europe, 2010.

Zmirou D, Gauvin S, Pin I, Momas I, Sahraoui F, et al. Traffic related air pollution and incidence of childhoold asthma: Results of the Vesta casecontrol study. J Epidemiol Commun. Health, 2004; 58-18-23.

8. ANNEXES

Table 10: Lagrange spatial regression - children's group - neighborhood matrix 1ª order.

CHILDREN / LAG - 1V				
Residuals:				
Min	1Q	Median	3Q	Max
-0.0800818	-0.0186020	-0.0104217	0.0047576	1.5434223
	Coefficien	ts: (asymptotic standard errors)		
	Estimate	Std. Error	z value	Pr(>|z|)
(Intercept)	1.5600e-02	8.8559e-03	-1.7615	0.0781525
DV_TOTAL	1.3131e-04	5.7269e-05	2.2928	0.0218569
MHDI_N_GRI	4.7456e-02	1.2535e-02	3.7857	0.0001532
	Rho: 0.17805, LR test value: 47.471, p-value: 5.5825e-12			
Asymptotic standard error: 0.025328				
z-value: 7.0299, p-value: 2.0675e-12				
Wald statistic: 49.419, p-value: 2.0675e-12				
Log likelihood: 6988.812 for lag model				
ML residual variance (sigma squared): 0.0031323, (sigma: 0.055967)				
Number of observations: 4781				
Number of parameters estimated: 5				
AIC: -13968, (AIC for lm: -13922)				
LM test for residual autocorrelation				
test value: 13.711, p-value: 0.00021317				

Table 11: Lagrange spatial regression - children's group - neighborhood matrix 2ª order.

CHILDREN / LAG - 2V				
Residuals:				
Min	1Q	Median	3Q	Max
-0.0580377	-0.0191446	-0.0105166	0.0048288	1.5465057
	Coefficien	ts: (asymptotic standard errors)		
	Estimate	Std. Error	z value	Pr(>|z|)
(Intercept)	-1.4800e-02	8.8943e-03	-1.6639	0.0961242
DV_TOTAL	1.3294e-04	5.7478e-05	2.3128	0.0207312
MHDI_N_GRI	4.4884e-02	1.2684e-02	3.5387	0.0004021
Rho: 0.2198, LR test value: 28.477, p-value: 9.4814e-08				
Asymptotic standard error: 0.040446				
z-value: 5.4345, p-value: 5.4958e-08				
Wald statistic: 29.534, p-value: 5.4958e-08				
Log likelihood: 6979.315 for lag model				
ML residual variance (sigma squared): 0.0031512, (sigma: 0.056135)				
Number of observations: 4781				
Number of parameters estimated: 5				
AIC: -13949, (AIC for lm: -13922)				
LM test for residual autocorrelation				
test value: 8.5581, p-value: 0.0034399				

Table 12: Lagrange spatial regression - elderly group - neighborhood matrix 1ª order.

SENIORS / LAG - 1V				
Residuals:				
Min	1Q	Median	3Q	Max
-0.06356949	-0.01766104	-0.01026061	0.00012829	1.42918326
	Coefficient	s: (asymptotic standard errors)		
	Estimate	Std. Error	z value	Pr(>|z|)
(Intercept)	4.2831e-02	1.1237e-02	3.8117	0.0001380
DV_TOTAL	-9.0095e-05	7.0335e-05	-1.2809	0.2002155
MHDI_N_GRI	-3.7265e-02	1.5726e-02	-2.3696	0.0178076
Rho: 0.1169, LR test value: 19.703, p-value: 9.048e-06				
Asymptotic standard error: 0.026176				
z-value: 4.466, p-value: 7.9694e-06				
Wald statistic: 19.945, p-value: 7.9694e-06				
Log likelihood: 5944.44 for lag model				
ML residual variance (sigma squared): 0.0048535, (sigma: 0.069667)				
Number of observations: 4778				
Number of parameters estimated: 5				
AIC: -11879, (AIC for lm: -11861)				
LM test for residual autocorrelation				
test value: 6.0922, p-value: 0.013578				

Table 13: Lagrange spatial regression - elderly group - neighborhood matrix 2ª order.

ELDERLY / LAG - 2V				
Residuals:				
Min	1Q	Median	3Q	Max
-0.06356949	-0.01766104	-0.01026061	0.00012829	1.42918326
	Coefficient	s: (asymptotic standard errors)		
	Estimate	Std. Error	z value	Pr(>|z|)
(Intercept)	4.2831e-02	1.1237e-02	3.8117	0.0001380
DV_TOTAL	-9.0095e-05	7.0335e-05	-1.2809	0.2002155
MHDI_N_GRI	-3.7265e-02	1.5726e-02	-2.3696	0.0178076
Rho: 0.1169, LR test value: 19.703, p-value: 9.048e-06				
Asymptotic standard error: 0.026176				
z-value: 4.466, p-value: 7.9694e-06				
Wald statistic: 19.945, p-value: 7.9694e-06				
Log likelihood: 5944.44 for lag model				
ML residual variance (sigma squared): 0.0048535, (sigma: 0.069667)				
Number of observations: 4778				
Number of parameters estimated: 5				
AIC: -11879, (AIC for lm: -11861)				
LM test for residual autocorrelation				
test value: 6.0922, p-value: 0.013578				

Atmospheric pollution - Respiratory morbidity pyramid

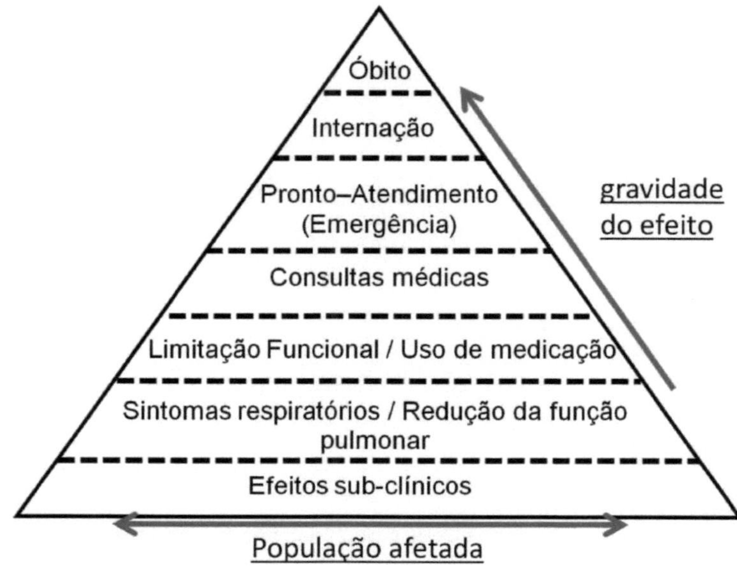

Source: Adapted from WHO (1999)

Figure 12: WHO - Respiratory morbidity pyramid.

yes

I want morebooks!

Buy your books fast and straightforward online - at one of world's fastest growing online book stores! Environmentally sound due to Print-on-Demand technologies.

Buy your books online at
www.morebooks.shop

Kaufen Sie Ihre Bücher schnell und unkompliziert online – auf einer der am schnellsten wachsenden Buchhandelsplattformen weltweit! Dank Print-On-Demand umwelt- und ressourcenschonend produzi ert.

Bücher schneller online kaufen
www.morebooks.shop

info@omniscriptum.com
www.omniscriptum.com

Printed by Books on Demand GmbH, Norderstedt / Germany